Apple Watch Secrets

Your ULTIMATE Guide to Getting the Most Out of Your Apple Watch

Rodger Peck

complete information. No warranties of any kind are declared or implied. Readers acknowledge that the author is not engaging in the rendering of legal, financial, medical or professional advice. The content within this book has been derived from various sources. Please consult a licensed professional before attempting any techniques outlined in this book.

By reading this document, the reader agrees that under no circumstances are is the author responsible for any losses, direct or indirect, which are incurred as a result of the use of information contained within this document, including, but not limited to, —errors, omissions, or inaccuracies.

Table of Contents

Introduction

We tend to see the Apple Watch as just a watch because it straps on our wrist and well, tells the time. That is typically the first mistake we make which blinds us from really looking at it for what it actually is – a wearable communication device.

The watch is no more a watch than your iPhone is a phone – a device merely to make a phone call. They are both so much more, and, with a little imagination, will allow you to do so much more with your life.

The watch enters the imagination at the crossroads of style and functionality in a way that no other watch or device can. Whether it's your Rolex or your iPhone, there is a niche that the watch occupies that you shouldn't do without. With that in mind, this book will exhaustively look at all the features the watch has to offer, both the ones that are typically known and the one that is secretly hidden in plain sight from the casual user.

In short, the idea behind this narrative is to take you from being a casual user to becoming a power user. If, however, you don't have an Apple Watch yet, then this is going to show you all the features on the surface and tech under the hood that will make you want to run out and get the watch.

Let's redefine your technology exposure. You have lots of reasons to get online or get to a laptop, but just under half of them are based on communication. These are things like emails, social media, messaging, all of which have one thing in common: communication and staying in touch. You don't really need a laptop for that unless you are still in 1999. The iPhone precluded the need to unfold your laptop, log in and check mail, it gave you the power of communication in the palm of your hands.

The watch takes it one step further, it puts all those little text messages, short messages, emails, social media updates, and even breaking news right there on your wrist without having to pull your phone out. In fact, you can use the watch to make and receive calls. Yes, that's right, now you can seriously leave your iPhone at home when you go to the gym. While you're there, the watch will be able to keep tabs on your heart rate, stream music and keep you up-to-date on messages. Two extra things to think about when you take the watch with you to the gym – you can have a sporty wristband for the occasion and you can pair it to your AirPods (or other Bluetooth earphones) for a connected, yet wireless workout.

We will look at all these features and more as we navigate the chapters of this book.

One of the things that slips the minds of most watch users is that it is compatible with LTE now which means that you get superfast internet speeds and that makes using Siri a delight. Which brings me to my next favorite feature in all this – voice assistance. With Siri, you can now get most of the assistance you need, and could once get from your iPhone, directly from your watch.

This brings me back to the point I made earlier of how the iPhone, a laptop, and a watch tied together allowed you to untether yourself from a bulky item to something that you can snap on to your wrist and forget that it is there. You can even go swimming with it since it is waterproof down to 165 feet. On any given day I don't see myself or most of you going deeper than 165 feet. Most scuba divers call it quits at 140 feet. Your watch will make it there and still keep an eye on your heart rate.

So, we can see that the issue of communication and social connection are definitely at an advantage with the data streaming to your wrists. We see, of course, that there are entertainment options in listening to music with paired AirPods streaming music, and then there is the personal butler you have at the end of a voice command, but there is still more – you no longer need to pull your wallet out to make a payment. Apple pay, which was enabled on all recent iPhones, is now extended to the watch. Just bring it towards the sensor and you are good to go.

For most watch users, the aggregate of all its ability means one thing, that they can use it exclusively when they are on the move without the need for a laptop, tablet, or, in some cases, even their phone. They could go swimming, for instance, and still be in touch with their world. They could go running and still get GPS and map input, or connect it to health apps to monitor their performance. The watch even has barometric pressure sensors so that an ascent or descent is recorded and the watch can figure if you are climbing or descending and at what rate; it goes towards a better workout analysis.

If you are out for a run and the music has kept you entertained, the health monitor has kept you aware of your performance, the GPS has given you directions, and you've been in touch with family and friends, there is still more that you can do. The watch even connects to HomeKit, which is Apple's home automation suitc.

Whatever your reason for finally making the decision to purchase the watch, here is more than you can do with it. From being a fashion statement and fashion accessory to being the quintessential tool that facilitates freedom from larger devices, the watch liberates your lifestyle and keeps you connected to the apps, services, connections and conveniences that you have grown accustomed to and need. If you have just purchased the watch, it will soon occur to you that it is a tool that is indispensable.

Chapter 1:

Basic Features

If you have an older watch and are still on the fence about switching over, there are only three reasons you need to make that decision. The first is that the watch is LTE equipped and that means you are not zipping much faster on the internet which will, in turn, allow a lot more services and apps to function smoothly. More importantly, in the past, your old watch had to be hooked up to your iPhone to get its online functionality, but that was the old watch.

The new one is all grown up and can access the internet, meaning it can get messages, social media, maps, apps and even Siri without needing to connect to an iPhone first or hunting for a Wi-Fi signal. The antenna for the LTE is located on the watch face, which is the most obvious and efficient location to place it.

The always-on LTE feature links to something called an e-SIM. Instead of sliding a physical SIM card like you would your phone, there is a chip that records the relevant information and mimics your SIM card that sits on your phone. In essence, you now have two 'phones' with the same number. That is what makes it more convenient. Personally, I never warmed up to the idea of having a separate number for my wearable device. With the watch that has the same number as my phone, I can even move from the watch to my phone seamlessly, mid-call.

The next thing you need to know is that Siri is totally independent on the Watch 3, and you don't even need to look at the screen when you are looking for a response. Siri will speak the response to you through the speakers on the watch, or it will transmit them to the AirPods in your ear. You will have to agree that this is about as Star Trek as it gets. This alone probably warrants an upgrade.

In terms of resolution and display, nothing has changed. You still get the same 38 mm option or the 42 mm one with 272x340 pixel resolution and 312x390 pixels respectively with the same 1000 nits display. If you are a tech newbie and don't understand what nits are, don't fret. Nits are the measure of how bright a display can get in relation to its ambient light. If you are out on a bright sunny day and have something that is low on nits, then you will need to shade your device to see what's on display, but when you increase the brightness of certain elements, the resulting contrast allows you to see the image clearly. A thousand nits are quite sufficient for you to be able to read your watch display in the brightest sun.

Remember that both the 2 and the 3 offer touchscreen interface so you can tap on it as you would your iPhone or other touch-enabled devices but, in the 3 as with the 2, you have a rotating crown on the side. This is a great feature in both models that allows you to interact with the information without totally blocking it out with your finger.

Aesthetically, the 3 and the 2 have little difference. They are both the same rectangular version and this gives you the best form factor for app interaction, but the three comes in an additional ceramic shell. You still have all the color choices from the 2 to be able to match it with your iPhone or iPad. There are also a number of new strap designs as well. If you have the 2 and are upgrading to the 3, the straps will fit both.

Under the hood, there are drastic improvements as well, specifically as far as battery performance is concerned. As devices take on higher capabilities in terms of connectivity speed and processing power, they also inevitably need more power. Batteries can only get so big before they become cumbersome, so what is needed is more efficient batteries.

The jump from Watch 2 to 3 saw a huge increase in battery efficiency allowing the 3 to increase its performance by 70% without increasing the apparent size. Additionally, if you step away from LTE and get to Wi-Fi, then that efficiency jumps to 85%. This makes a huge difference between the 2 and 3 in terms of how long between charges you can stretch the watch.

The accelerometer that was in the 2 is also in the 3. Between the GPS and this accelerometer, the 2 was able to give you good running information, but with the 3, that is kicked one notch up with the barometer being able to detect elevation change as well. It is sensitive enough to be able to tell when you are climbing a flight of stairs or jogging up a mild slope. With the right app, your workout and progress gets granular and you will have better control over your session. It is best used to map out paths and to change your running or biking routine. Using it on Black Diamond ski slopes or mountain bike paths is an interesting way to use it as it builds the ride profile for future reference. The difference between the 2 and 3 is that to get altimeter information in the 2, you needed your iPhone close by, while, in the 3, that is no longer necessary.

With the right App, like GymKit, you will also be able to interface with the equipment at your gym. This means you can share heart rate data, height and weight with the equipment and tailor your workout. That, in turn, can be recorded and you can get a full profile of your outline for analysis and planning.

As you made your way through this chapter, the thing that would have begun to become obvious is that the capability of the iPhone, or any smartphone for that matter, has moved gradually and efficiently to your wrist, while the capabilities for the phone and its larger screens have started to take on the functionality of a laptop or tablet. That makes the watch, more so than other smartwatches, an imperative inclusion in your tech arsenal. It's like the decision to jump to a smartphone ten years ago liberated you from the laptop, and the move from the desktop to the laptop twenty years ago untethered you from the desk. The watch is no longer a nice-to-have; it has most certainly become a must-have.

Chapter 2:

Setting Up Your Apple Watch

The first thing you should do is to pair your watch and your iPhone – if you have it. If you don't, just move on to the next section. Pairing your iPhone to your watch simplifies a number of matters and, more importantly, gets your watch to act as a surrogate phone. This is when the watch copies the SIM information and is able to go online and accept incoming communications.

Start by making sure that your iPhone is on, active and is connected to the internet through cellular or Wi-Fi. Go to Settings>Bluetooth and activate Bluetooth. Once these are ready, it is time to get the watch set up.

Put your watch on, and then turn it on by pressing and holding the button on the lower right side of the watch. If you just bought it, the battery should be fully charged. Keep it pressed until you see the familiar Apple logo appear and then release the button. Close the distance between the watch and the iPhone. You will see a notification on your iPhone "Use your iPhone to set up your Apple Watch." You can tap 'continue' and, from here on in, keep the phone and the watch in close proximity until the set-up is complete.

The watch will soon display a 3D cloud in motion. Place your iPhone camera over it and focus on it so that the camera can capture this event. The phone will then ask which wrist you plan on using the watch, so chose your left or right and continue. The Apple terms and conditions will appear, and you now need to accept them by tapping to agree and then tapping 'continue.'

At this point, it's pretty much menu-driven, and it gives you the opportunity to set up your Apple ID and to possibly restore your Apple Watch 3 from all that you had in Apple Watch 2. If you plan on using Apple Pay, then you will also have to set up a passcode to get into your watch. From there, you just need to give it a few moments for the watch to sync the two devices. At that point, you are ready to use your watch.

Where Do We Go From Here?

At this point, you are in one of two camps. Either you are thinking to yourself "What else?" Or you are thinking, "There's more?" Either way, it would surprise you as it does most people that what we've covered here is just scratching the surface. We may have just looked at 10% of what you can do with the latest watch.

There are three areas that set this watch up as an essential wearable device: connectivity, app library, and interface. We will look at each in turn and you will be able to get to know a number of features and hacks that will make your watch indispensable.

Connectivity

This is not just about getting online. It's about all kinds of connections. From phone calls to text messages; from social media updates and interactions to email and encrypted communication - everything that you can do to connect to the world at large on your phone or laptop, you can now do with a device that is attached to your wrist inconspicuously.

Let me explain. To make a phone call on your iPhone, you need to pull it out of your pocket, get past the lock screen and invoke the phone function. From there, you can either get Siri to make the call, or you can scroll through your contacts and connect from there. On the watch, you just tap your wrist and tell Siri, which is always on, to call your contact. The Siri environment on the watch is extremely conducive to messaging.

To read an incoming message on your watch, all you have to do is raise your arm and the message will be displayed on the screen. You can dismiss it by swiping it. You can do the same with a phone call. You do not need to activate the display as it will light up as soon as you raise your arm, and you can choose to answer the call or decline it. But what is really neat about the watch is that you can send canned responses to incoming messages. Just respond and select preset messages and send the response you want, for example, "Running late," "Call you back" or whatever you want.

The best way is to save all the most used messages and invoke them when sending a message. It's even easier if you are initiating the call, just use Siri. "Hey

Siri, message mom. Tell her I will be late." It's that simple. Or you can record an audio file and send that as an attachment. Simple, efficient and classy.

You can also control how you are notified of incoming messages or alerts. There are, of course, the audible notifications but in the event you want it to be less intrusive, or you wear it on your wrist while you sleep, there is a haptic engine that will alert you via a sensation on your wrist. The haptic response is silent and only you can feel it because it interacts with the tactile sensors on your skin beneath the watch.

Being alerted to events and alerts this way allows your communicates to be discrete amidst live interactions with others.

This brings me to the eight tips you need to set up your Apple Watch when you first get it so that all your notifications are exactly the way you want it.

Notification Set-Up

There are two ways you can set up aspects of your iWatch. You can access the app that connects it to your iPhone (the Apple Watch app) and then manage it from there, or you can manage it directly from the watch itself. Most people typically use their iPhone to set up the watch, unless they are doing it on the fly.

On your iPhone's Apple Watch app, just head to My Watch>Notifications and, from there, you can set up which app, service or alert will activate the notification on your watch so, for instance, you can disallow active Whatsapp notifications and, instead, you will receive the message but it won't alert you to it. You will have to consciously check the watch for any incoming messages. On the other hand, you can switch all incoming messages to silent when you go to sleep and only allow emergency messages to alert you. All this can be programmed on your iPhone or on your watch.

But here is a trick! I bet you already have your iPhone set up exactly the way you want it. If so, then all you have to do is scroll down the Apple Watch app in Notifications and you will find the option to Mirror your iPhone. Select this and you will find that your watch will notify you of incoming alerts and messages in the same way your iPhone does. It's a good way to start things off. You can then alter them individually to fine-tune your preferences.

It is highly recommended that you set up all your notification preferences as soon as you pair up your phone and your watch.

The one thing that confuses new watch users comes from the fact that you are able to get notifications, incoming messages and calls on either device. It can't go to both. You can't have both devices ring when a call comes in, and you won't have messages appear on both devices. They could be synced together later to put all the incoming and outgoing messages in one place, but they can't be done simultaneously. Two things to note here is that when your phone is asleep or locked, all calls automatically go to your watch. As far as apps and notifications, it depends on what you set up in the app. I typically lock my iPhone X and leave it at home when I go for my run. All my calls and messages re-route to my watch. Easy.

The key to how and why the watch works is the fact that the powerful processors and fast connectivity allow for some highly effective AI technologies to be at your beck and call right there on

your wrist.

Chapter 3:

Ten Secrets of Force Touch

Force Touch brings an added dimension to tactile instructions. It first appeared in the larger Apple devices when it was not just limited to swiping to invoke an action, but a forced touch on a spot, similar to clicking but without the dip and spring back. Force Touch brings a whole new dimension of functionality to the watch, espccially since so much needs to happen in such a small footprint.

To check your messages, just swipe down on the screen and then press on the screen to clear all the messages. It may take some getting used to. The swipe down is especially for the notifications that you have not set up in your app. It won't notify you of those messages that you choose.

For instance, I have notifications only for VIP emails. Any other email that comes in will not trigger a notification. After all, I have other things to do, but when I have time and want to check those other emails; I activate my watch, then swipe down as mentioned. Now I get to see all my alerts and messages that I was not notified about. I can scroll through them by swiping up with a finger, or swipe down, or I can use the crown and roll them up or down. From there, if I want to clear them, I just have to press down on the face. Easy since it doesn't have to go through the numerous steps of menu-driven actions.

Force Touch can also be used to bring up a menu on the watch face. So, let's say you want to change your display and its contents, just force touch it to invoke the menu and you can change the face and the elements of it, including how it is arranged and the various themes that are included.

The third secret to Force Touch is to use it when you are in Messenger. If you are reading a message, the Force Touch will move you straight to composing a reply or, if you are not in a message, Force Touch will take you to a new message composition and allow you to choose the recipient(s) from your contact list. You can even use Force Touch here to share your location from this location.

Another secret of Force Touch that you won't typically find in the manual is that you can use it to close the Workout app when you're done and it will save your data for review later. You can also use it in the Activity app which will then allow you to change your daily goals for the amount of activity you want to accomplish. I use this on a daily basis, especially when my world demands me to be at my desk for a large part of my day. It makes it fairly easy to forget to move. The Move goal allows me to keep track of how much movement my body is getting. You can invoke the Move target with Force Touch.

That's five. For the sixth, you can use Force Touch while you are on a map and then from there either share your location, as you would in the messaging app, or you could search for an address right on the map app.

The seventh trick with Force Touch is how you handle your music. Instead of fiddling around the menus and getting around the music app, once there you can Force Touch to choose your music, its source – whether it is from your iPhone or the music that is local on the watch, and you can choose to send it to your AirPods or Bluetooth speakers.

The eighth trick is the way you get your alarm on the watch. You can Force

Touch within the Alarm app and it will allow you set an alarm right from the watch itself.

The ninth is how you use Twitter with Force Touch. Within the Twitter app, just press hard and you can compose a tweet from there.

Finally, the tenth, you can use Force Touch on your calendar as well. It will take you to your day's events or your task list.

These are just ten of the things that you can use, but if you remember the rationale behind Force Touch, then you will be able to find tons of other uses as more third-party apps are starting to take advantage of the added functionality and ease of access.

Chapter 4:

Power User's Guide to Extending Battery Life

There are power users, and then there are POWER users. How you use your Apple Watch will differ widely between one person and the next because there really is just so many things that you can do with it. If you predominantly use your iPhone and have the watch as just an accessory, then your battery life is going to extend the farthest. But that's not why you bought this watch, is it? You bought it so that you could do more, with less. You bought it so that you can leave your iPhone during times that it's not convenient to carry it, or you need the iPhone to charge. As such, the tricks you will find here are designed for you to extend the use of the device and to prolong its interval between charges simultaneously.

To get started, let's look at how to get into the control page. You do that by swiping up from the bottom of the screen. Now you are in the Control Center. Here you will see the battery status on the top left corner. If you tap on the battery, it will expand to the full screen to show you the state of the battery's charge. It also puts up a slider for you to activate the battery saving mode. This is one feature that you should take advantage of, especially if you are hesitant to keep charging your phone often.

There are two things that suck up the power on your Watch 3. The first is the cellular service, and the second is the number of apps you run on the watch. The latter you can control, the former you have to keep an eye on. Here is how:

The watch will switch to cellular service automatically if you keep it away from the iPhone. If they are paired, then it takes its connectivity from the phone, if it is not, then it goes straight on the hunt for cellular service. The good news is that while the first generation could last 18 hours, the third can last up to 38 hours with fairly constant usage.

The next thing you want to do to stretch the battery life is to use the least active watch face. If you have it fully animated, then the watch is going to suck down juice heavily. Instead, use a static watch face with as much black as possible.

Here is something that you need to know - the battery that is in the watch does not need to be totally discharged before it is charged up, so you are indeed free to charge it on a nightly basis. It automatically stops the charging when the battery is full and it will not diminish the useful life of the battery. It will even restart the charging as the power is used through the night. The smart charging gives you the option to leave the watch on a charger if you plan to be away for a few days and are taking your second watch with you instead.

Go ahead and put it on the charger nightly, even if there is still a charge left in the battery. It's easy enough to place it on nightstand mode while it is charging that way, and you can use it as a night clock while it charges. The digits of the clock will even re-orientate so that it is facing the right direction as you lay it on its side.

To really squeeze the juice out of it, here are a few more things you can do. Go to General>Accessibility>Reduce Motion and turn it on. This will reduce the watch's design to constantly re-size or animate the watch face when an app is called.

Between this and the static black face, you will find that there is a significant saving in battery life, and there is more.

The next item you want to turn off if you want to reduce power consumption, especially if you are the kind of person that is constantly raising your hand, is the Wrist Raise. To turn this off, use your iPhone, then go to the watch app and tap General>Wake Screen and turn it off. When you do this, you will find that many of the automatic functions that come with raising your arm will stop. Even some of the activity sensors and measurements will cease to record, but you can adjust these individually as well. You will just have to turn them on individually when you need to use them. It is not practical, but it will save you a load of power.

Another potentially huge battery saver is notifications. Turn these off so that you are not alerted every time a message comes in. It's also a great way to not be distracted. Another related way is to stop touching others. It's called a Digital Touch, or a tap, you know that you can send a touch to someone using your watch so long as they have one and they are wearing it, right? The haptic sensor that is used to alert you is also used as a touch on their end. If you want to save power, don't use this. It's a good way to send someone a nudge without having to send them a message.

If you go running or biking once a day, wear it to work and keep it on all day, and you would have your activity tracker on, as well as Messenger. That would not be considered power usage and you could possibly get away with charging it once every two days. However, to be able to use it more frequently than that, keep it close to your iPhone as much as possible and, when you separate the two, then keep the apps that are running to a minimum. That way, it is drawing power for its cellular connectivity, but it is not running the power to keep the apps in active memory.

If you feel that there is more you can do to cut down the power consumption, then head over to the My Watch app on your iPhone and then look at all the apps that are active and it will tell you how much power each app is drawing. Just go to the app, type General>Usage to find this list. When you know what is necessary, what is not, and which one is draining the most, you can force quit that app.

To force quit an app is a little different

from an iOS device. You have to open the app, and then press the power button that is on the right bottom side of the watch. Once the face displays the power down, press the crown that is on the top left side of the watch. Press and hold until the app shuts down. You only have to do this a few times and you will know which app sucks up more power.

Another way to save power is to remove the apps that you don't need. The memory in the watch is maintained by the use of power. The more apps that are installed, the more power the watch will use. For this reason, if you are not using the app, remove it and you will see there is a perceptible difference in the longevity of the battery.

Finally, there are three things that you can do that most people won't even think of that will significantly cut down on power usage. Switch to Airplane Mode. This will cease all transmissions and receive signals, including Wi-Fi and cellular, but Bluetooth will remain. You really can't turn this one off or else you will end up not having any functionality. It really becomes just a boring watch at that point. You can then turn the display to grayscale instead of color and dim the brightness.

The last few power-saving tips are radical and are included here in times when you can't get to a charger and need the watch to stay operational for a few more hours. You can pull the life of the battery significantly that way.

The best way, however, is to keep the phone charged every night, all night and then use it freely.

Chapter 5:

Apps and Functionality

The heart of the watch is balanced by two aspects. The first is the hardware that we have looked at thus far. The second is the apps that reside on top of the operating system. Without the apps, the watch really is not going to be able to do much. There are native apps that reside on the watch when you unbox it. Above and beyond that, there are more than 4,000 apps in the app store.

The best way to think about the watch is to see the areas where it facilitates your existing or intended lifestyle. At its heart, the watch promises hands-free mobility with voice command assistance. It takes all the core connections you need and puts it on your wrist. As we said in the beginning, to really make full use of the Apple Watch, you have to stop thinking of it as a watch, and start to look at it is a wearable computer.

When you look at it as a computer, you then realize that the computer is only as good as the apps that run on it, and the amount of connectivity that it supports. On both those counts, the Apple Watch, especially the latest version, aces that. The watch comes with some native apps and they give you initial functionality out of the box, but there are numerous apps that can extend your effectiveness and you can check them out at the app store. What you do need to know, however, is how to download and install them.

To install an app on your Apple Watch, you first need to have the Apple Watch app installed on your iPhone. Once you have that, open the App Store tab that is located at the bottom of the screen. Within the App Store, locate the app you want to purchase. Once purchased, it will appear in your iPhone within the My Phone app. Look for it in the list and then move the slider to place the app on the phone. If you give it a few minutes, you will then see the app on your watch.

To remove it, an easier way than described earlier would be to go to the My Watch app on the iPhone and locate the app then push the slider in the opposite direction. Give it a few minutes and the watch will no longer contain the app.

The earlier method of using the On button and then the crown is for times when you do not have the iPhone with you.

Chapter 6:

Camera

While there are no cameras on the Apple Watch, there are camera apps that will make your remote image capture easier to accomplish. There are a number of apps that will facilitate this in addition to the native Apple apps that will make viewing images on your watch possible.

The addition of a camera on the watch would have been rather cumbersome and ridiculous. Imagine having to contort your arm so that you could point the camera towards the object of the intended image. Even raising your arm up to have FaceTime with video seems a little awkward. No. The use of the watch with an image capture device is not intended because it just doesn't make sense.

However, when you pair the iPhone to the watch and you start thinking of selfies, group photos, remote image capture, then what you should really remember is that the primary camera in the back, the one not on the same side as the display, is the one that is superior.

Even with selfie sticks, you are faced with two crippling problems. One is that the stock only goes a limited distance out from where you stand, and that results in two issues. The first issue is that one of you, most likely the one who owns the iPhone, is going to be captured in the image as the one with his or her arm stretched out awkwardly. That's a small issue compared to the fact that a selfie stick only gets you so far and then you are unable to fit everyone in the picture, or you have to hold the camera up in the air. The watch precludes all that because you can now view the frame that your iPhone is focusing on. Here is the beauty in that – you can face the powerful iPhone camera that comes in the back towards your target. By doing this, you get a better quality capture, and because you can place it as far as you want, you can get the whole group into the frame. You can't move the picture or adjust the composition, but you can place the phone on a tripod then adjust it all you want, get back to the group and check your watch for composition then click the shutter from your watch.

You don't even need to activate your iPhone. You can just open the Camera app on the watch and that will wake your phone and turn the camera on –

the phone remains locked. You can then view the image on the display and tap what you want to focus on. There are three tricks that most users miss.

The first is that you can give depth to your image by tapping to focus the foreground by tapping closer to the bottom of your watch. That brings what is nearer into focus and leaves the rear slightly out of focus. If you choose the back, then the front goes a little out of focus. Once you've chosen the focus point, just tap the shutter button. That's the first trick.

The second trick is that you can use the

timer button, which gives you three seconds, to take rapid shots. If you just tap the timer, as mentioned earlier in the chapter, then you get one shot at the end of three seconds. If, however, you keep pressing the delay button on your watch, the camera on the phone will take consecutive shots at about 5 frames per second, but this is not so that you have multiple frames of the same shot, this is so that you have the best shot to choose from later.

The third trick is to use Force Touch together with the camera app. With the app open, Force Touch the display and a secret menu appears. From here, you can control the HDR function of your phone's camera, you can activate the flash, you can involve Live Photos and you can even flip the camera from back-facing to the front-facing camera.

The neat thing that I like about this is if you have you haptic on. Each time you press the shutter button, it will send a sensation to your wrist like a tap to tell you the shutter was released. Neat. By the way, you can also use the power button as a shutter release button.

Chapter 7:

Hidden Secrets

There are many hidden secrets to find and paths to take within your watch environment. It can range from just neat to do, to important and time-saving. There are many more shortcuts and hidden surprises in the watch than is listed here, and each subsequent watch OS version will bring more, but once you get through this list, you will get an idea of how to navigate your way around the watch to be able to find others. Swipe methods in different locations, Force

Touch in different displays, crown pressing and scrolling, and the power button result in a number of surprises that you can play around with. These are just the ones that will make you a power user. You can either randomly explore all your watch has to offer, or we can help reveal some of the more useful secrets that are built into the watch. Here we go:

Find the power button on the lower right side and press it twice in rapid succession. This will bring up Apple Pay whether or not you have set it up. If you have set it up, you can then make your payment from this point.

Right above the power button is the crown. This is the feature you use to scroll through menus. You can also tap on it. If you double tap on the crown, it will take you back to the last app and the place you were at.

You can even take screenshots on your watch. Whatever is on your screen, just press the crown and the power button screenshot – press crown and power button simultaneously. This will take a screenshot and save it in your Photos folder on your iPhone.

There are times when your watch can

become unresponsive; it is just a fact of life. All technical stuff will do that. Not to worry, just press the crown and power button simultaneously for about ten seconds. That will reset and bring up the power on the screen.

If it's just the app that is hung or you need to shut it down for whatever reason, then you can force close the app. To do this, from within the unresponsive app press the power button until the shutdown menu emerges then press the power button again. The app will shut down. You can start it back up fresh again.

To delete an app, press and hold the app page (don't press too hard or it will activate the Force Touch). Then click the x on the app. Once pressed, it will ask for confirmation. Deleting it on your watch will not delete it on the iPhone.

You can clear all the messages in your notifications menu in one click without having to do them individually – that can be a pain. SO, here is how you make it simple. Open the notification window then press a message for a moment and a 'Clear All' message pops up. Just tap on that.

Many of us refer to the watch for

weather and forecasts but, to do this, you need to set the default city and you can only do that on your iPhone. The Weather app will then take the iPhone's current location and use that, or you can choose another city from the menu. When you are getting ready to go to a new city, it is a good way to get a forecast. On the weather display seen, if you tap it, you will get the temperature forecast, and if you tap it again, you will see the probability of rain. You can also Force Touch the screen in the Weather app. If you do, you get the option of temperature, probability of rain and weather selection so that you can choose which one you want instead of sequential tapping to cycle through the selection.

The next quickie move you can make is in the clock view. If you swipe up from the bottom of the screen, you get to Glances. From here, there are four things you can choose from. One of them is the ability to ping your iPhone. If you forget where you left your phone, or if it fell between cushions, all you have to do is ping it from your watch. But there is more. If you hold down the ping button on your watch, not only will it ping it, but it will also repeatedly illuminate the flash on your iPhone so in the dark, you will be able to find your phone easily.

If there is something on your screen that may be a little small for you to focus on, you can zoom in. To do this, go to your Settings page on the watch, and then go to General>Accessibility and then activate the zoom from there. Once you do, you can zoom any display that you have on your display by triple-clicking the crown.

While you're at it, while you are in accessibility you can also activate VoiceOver.

Under Accessibility, you can also adjust the strength of the Haptics feature, whether you want to strike your wrist gently or with a little zing to it. Even if at full setting you find that the Haptic could be a little stronger, then head on over to the iPhone and get into My Watch>Sounds and Haptics and select Prominent Haptics. That will give you a significant tick when an alert is given.

Many users think they should only use one input method at a time, but, actually, you will move faster and with more precision if you use two different modes to manipulate your watch. For instance, you could raise your watch up and say "Hey Siri" and that will get Siri up and running for you, then tell it to Launch Settings. This will launch the settings menu and, from there, you can use the crown to scroll to see what's available. Once you get to know your phone much better, you can tell Siri to do exactly what you want without just using it as a voice command to navigate the menus.

So, let's say you want to adjust the Haptics. You don't necd to move progressively through the menu, just call the Launch Settings and call for Sounds and Haptics. You can just tell Siri to raise Haptic and it will be done.

Another hidden secret that is handy is handing off your conversation or your session on the watch to your iPhone. To do that, just turn on the phone and, at the bottom corner, you will see a symbol of a watch, drag that up to the top of the phone and your passcode page appears. If you have a passcode, enter that now, or if you are on touch id, place your thumb on the button. If you are on an iPhone X, then your face would open the phone without any additional prompting.

There are times when you may be in a meeting or otherwise indisposed to take a call on your watch. When the call alert sounds, all you have to do is cover the watch with your palm, and that silences the ringing. It doesn't do anything to the call.

Most people think that when they raise their arm that the watch automatically brings up the clock face. That's true only to the extent that it is the default setting. You can actually change that.

Just go to Settings>General>Activate on Wrist Raise and choose whether you want the last app to come up or the watch face. You do this on your iPhone.

Another feature that is hidden from most users is one that every user passes by on a daily basis. When you receive a call, the watch displays the caller ID and the two buttons to answer or kill the call. We told you earlier that you can silent the alert of the incoming call by covering it with your palm. But there is more. If you swipe up from the bottom of the screen (not half way), it reveals a couple of other things you can do with the incoming call. You can choose to send it to your iPhone or you can send the caller a message.

While we are talking about messages, you can set default messages so that you can respond quickly without having to dictate a message or type one out. Tap on the Siri icon and choose a canned reply to send out a message.

A good way to customize what view you want in Calendar is to Force Touch into the calendar and you will be given a menu to choose if you want a day, week or month view. That makes it easy to move around your schedules and get a glance at free times.

When it comes to secrets about the watch, there are many ways you can check the battery levels on your watch. You can do it from within the watch, or you can do it from the My Watch app on your iPhone. The one thing that people never realize what they can do is to check the battery status of their iPhone on their watch. You can do that. Just invoke Siri and ask it, "What is the battery level on my iPhone?" Siri will return the percentage of the battery to you.

One more hidden secret that we stumbled upon is the time stamp on any message that comes in. If you just pull the message with a swipe to the left on top of the message that you are curious about, the message drags left and reveals the time stamp.

These are just some of the most popular hidden secrets in the watch. There are tons more and you can fiddle around the watch to get to them. There are also other secrets in other chapters of this book; in fact, this book is designed to be a trove of secrets in different areas. We just put them under different headings.

For some reason, Apple made the watch chock-full of functionality, but didn't document them all in a conventional manner. They left us to do the discovery, but they also meant it to be a kind of an adventure where you get to play around with your watch and fiddle with it as you get to know its 'ins and outs.' This book is just the beginning, and you will only get to know your watch intimately when you fiddle around 8with it.

Chapter 8:

The Digital Crown

It is widely assumed that the Digital Crown located on the upper right side of the watch is what Apple's home button is to its iPhone and iPads. It takes you back to a common location to get you started, but the crown is significantly more than that. It holds a number of secrets and, if you can master these secrets, it is going to prove that much more effective.

There are so many things you can do with the crown that it is hard to really just pick one, so we are just going to start with the easiest. Typing messages can be a pain so Apple uses its legendary auto-complete feature which, if you are used to knowing that, it can be cause for some pretty embarrassing tweets. You know, you type one thing, it puts in something else. To avoid that in the watch, while you are in the composition window, tap the crown and scroll through the word suggestion for your partially-typed word. That way you don't have to spend so much time trying to fore-finger your way through an email. You can also invoke this as you scribble letters on the face. For this, write the alphabet with your finger on the watch and after you get two to three letters of a word in, use the crown to scroll the suggestions. It just takes a little time getting used to it and, once you do, those times when you leave your iPhone at home, messaging becomes easy.

You can run more than one app at a time, and to do that you can use the crown to assist. All you have to do is tap once on the crown while you are in one app, and that will bring you back to the list of apps, then you can choose another app to run. One other way to think about the crown is to look at it as your go-to button when there are no other soft buttons on the screen. You can tap it once, twice or even three times to see what it does. From the second app, just double tap the crown and you will switch back to the other active app. This comes in handy if you are on the music app and workout app while you are running, for instance.

The crown in the app menu is also a really good way to zoom in. When you have your app cloud in the screen, just roll the crown and the icons zoom in and out smoothly. Once you get it to the size you want, just tap the screen and you get your app. The zoom in and out feature directly from the crown only works on the app page. For either zooming in or out, refer back to the earlier part of the book.

We've talked a lot about Siri, but many don't know how to launch Siri. This is where the crown comes in. Press and hold it for two seconds and Siri comes alive. The thing about Siri on your watch is that it has a bit of a hard time separating the ambient noise because the microphone on the watch does not have active noise canceling like the iPhone has. It will end up misunderstanding your commands so, if you are in a noisy area just raise your watch closer to you.

Chapter 9:

The Most Common Issues You May Encounter With Your Apple Watch and How to Fix Them

Any technology-based products from your car to your TV is going to have some kind of a problem along its lifespan in your possession. For me, those problems are either product related or user inability related. The second one we can fix by looking around for solutions, but the first one needs a little insight into how the device works.

For now, you have an idea of the inner workings of the Apple Watch, but we are going to take it a little further and show you how to troubleshoot, as well as get the watch to do the things that it initially is hesitant to do.

The fastest and easiest thing that you can do when something is not working is to restart the watch. In most cases, watchOS takes care of problems when it reboots and this typically gets everything back to order. There are three categories of this. You can either restart, force restart or reset your watch depending on the severity of the problem. To facilitate an effective restart, you should also restart your iPhone and let the two devices reconnect to each other as well.

If you can do them simultaneously, that's fine, if you can't, that's okay too. You should already know how to restart the watch based on earlier instructions. You should also already know how to restart your iPhone.

If your restart on the watch is not successful or it doesn't fix the problem, then do Force Restart and you do this by pressing the power button and the crown for about ten seconds, and when the slider appears, just slide it and the watch will restart. Do the same with the iPhone according to the iPhone procedures.

If none of that works or you have the need to reset the watch, then you need to go to your My Watch App on the iPhone and navigate to General>Reset and Erase All Content and Settings. You do not need to do it to your iPhone unless there is an iPhone problem and you should consult your iPhone's documentation for that. If you do this, be warned that it will delete everything that is on your watch and you will have to pair it to your phone again. If you have backed everything up on your iPhone, then doing a re-sync will take care of it.

If you forget your passcode, you should take note that it will behave in almost the same way as other Apple products. Upon failing your tenth try, it will erase all the data on it if you set the Erase Data option in your iPhone. You should do this especially if you have sensitive data on the iPhone, and even if you don't, because the iWatch can still be able to pair with your iPhone and that represents a security risk.

The first common problem that occurs among watch users is the inability to charge the watch. There are two ways you can troubleshoot this and, in most cases, that gets it fixed. The first is for you to reset your phone. The second (and this is more apt if this is your first charge after purchasing it) is that you need to check that the magnetic charger is bare. That means all wrapping and the plastic stocker that is placed on the charging plate during packing, is removed. There should be no material inhibiting the contact between the charger and the back of the watch. Make sure that the back plate on the watch is dry and there is no grime or dust on either surface. Once all that looks good, follow the cable with your hand and make sure there are no nicks and cuts along the cable and follow that to the USB connection and then insert the plug. If none of these are damaged, then plug it into a power outlet that you know works and try it again. If the charging symbol does not turn green after 30 seconds, then use a different charger, a different USB cable, a different adapter, or a completely different set that you know works. If this still doesn't work, do a reset of the watch and try it again. This usually gets it going. If it doesn't then you need to contact customer services.

Another highly annoying problem that some users face is the fact that the watch would not automatically handoff from Wi-Fi to LTE. If you have this problem, try a restart and that will usually take care of it but, if it doesn't, you need to make sure that you are running the latest watch OS. Just because you bought it this afternoon does not mean that the OS is up-to-date. You still need to update it. Once you update it, try it again. To update the OS go to your paired iPhone, get to My Watch app and head to General>Software Update. Most of the time the problems exist when the watch logs on to unauthenticated Wi-Fi networks.

If that still doesn't work, there are still a couple of things you can do. First, try to reset your cellular connection again. To do this, you need to forget your old connection then get back on again. Go to the iPhone and fire up the My Watch app, then go to the carrier and tap on the 'i' located right next to the name of your carrier. Once you are there, click Remove. Once it has been removed, now you can click on Add Your Service Provider again. Once you tap on

Reactivate, restart your watch. This should take care of it. If not, it is time to visit your Apple dealer.

The next on the list that may be an issue is the problem of not being able to connect to your iPhone. Sometimes, the problem may be as simple as the fact that the phone or watch may be on Airplane Mode. In that mode, the Bluetooth is automatically turned off so the two devices will not be able to pair. As such, your first task would be to check the Airplane Mode. If it is on, it wouldn't hurt to recycle it. If that doesn't work, go to the control center and confirm that the phone and the watch are paired. If they are, then unpair the two devices and pair them again. If none of these work, restart the watch and restart the iPhone.

One of the more common occurrences is the sudden inoperability of the Digital Crown. No matter what you do, it does not scroll or it does not register a tap. Your first workaround for this is a total reset. This is where you take the watch back to factory settings. We showed you how to do that earlier in the book but, before you do that, think about if this is a mechanical issue or a software issue. A software issue is one where the crown moves freely but nothing happens. A mechanical issue is when the crown is stuck or it feels like its grinding against something. In case it is the latter, then you will need to clean it according to Apple's instructions which are, essentially, to turn off your watch and place it under a faucet's stream of warm water. The water should flow over and around the corner and between the slight gap that exists between the crown and the chassis. If neither the mechanical cleaning nor the electronic reset work, then it needs to be sent back to Apple.

Another common problem that has plagued certain users is that the activity tracker doesn't track or is incorrectly tracking the activity. Before you consider zipping over to the dealer, there are a few things that you can try. Make sure that your Bluetooth is paired to your iPhone and connections are not a problem, and then go to the My Watch app and go to General>Wrist Detection. If it is off, turn it on, and then put on the watch if you aren't wearing it already. Make sure that it is firm, as there shouldn't be any gap between your skin and the back plate on the watch. If you need to, adjust the band and then check to see if it begins tracking your information. If it is still not, check your medical information and make sure that your information is accurate and exists in the database. Without height and weight information, it would be impossible for the watch to calculate the necessary data. If all this is done and the phone is tracking erroneously, then it possibly needs to be recalibrated. For this, you need to go to the My Watch app on the iPhone and go to Privacy>Motion & Fitness>Reset Calibration Data. That should get you up and running.

These are some of the most common

issues that you will come across when your watch doesn't function the way you expect it to. The fixes are simple and you just need to go through the motions to get them up and running again.

Chapter 10:

What's Coming Next

The Apple Watch is a fairly sophisticated piece of technology and it is Apple's foray into the wearable computing market. It certainly won't be the last. In the last four years, there have been three iterations of the Apple wearable. The first watch entered the market in 2015 and, frankly, wasn't the most impressive market leader at the time. That, however, has changed considerably today. The third version of the Apple Watch, which was released in 2017, has everything you need to be able to leave your iPhone at home. You can go out on a run while still having much of the resources you need to keep in touch via voice, text, social media, as well as be able to consume entertainment and keep track of activities.

Wearable computing is really at its nascent stages and Apple has undoubtedly leapfrogged to the head of the class among its competitors that have tried to bring in other features.

However, as good as the current version is, the rumble on the tracks seem to suggest that the best is certainly yet to come. It has been credibly suggested, according to patent filings, that the display footprint is going to creep outside the four corners of the watch face and move into the band as well. After all, there is a large amount of real estate along the perimeter of one's wrist.

What many have asked for, and it looks like Apple is listening, is a circular face. If that's what you've been waiting for, then it's time to keep a look out because it should be in the offing in 2019 or 2020. If the watch does go circular, then the band having extra sensors and display will make it a highly functional upgrade to an already awesome wearable computer.

Looking further out beyond 2020, Apple's trajectory suggests that some form of holographic projection will be included to be able to project a full-size keyboard onto a surface, as well as the possibility of a camera in the watch to support FaceTime with video. The camera would not necessarily be used for capturing stills and video, but it would be a good option for communication.

The future of wearable computing rests on three areas. Firstly, it is about battery and charging technologies. Secondly, it is about processing and memory which is getting smaller and denser over time. The possibility of quantum processing and quantum storage will certainly be a boost for wearable computing, especially something as small as a watch. More space can be allocated to display and less to hardware. It will also end up being lighter and more powerful for any given unit of size.

Finally, the future of the Apple Watch looks like it will include increased functionality and larger units, while being marginally heavier. The idea, it seems, would be to move as much as possible away from the iPhone and into the watch, while increasing the capability of the iPhone.

Conclusion

If you already own an Apple Watch, then this book would effectively raise your level of interaction with it. If you are yet to get one, then I congratulate you on your intention of getting the watch. If you have held out since the first model, then this is the right time to get it as the technology, price, and the features are optimized at this level. You are going to get the best technological bang for the buck.

As mentioned at the beginning of the book, just because it is called a watch, doesn't mean you should look at it as one. There are only two things it has in common with a watch. The first is where you place it (your wrist) and the second is that it tells you the time, but, after that, all its other features go far beyond what you could rationally label as a watch.

Its design is to give you all the basic functionality of your iPhone. Many of the apps that you use to drive your phone are available for your watch as well and, while you can run the watch independently, it really is best if you think of it as an extension to your watch.

This book is designed to do two things and that doesn't include bringing you up to speed on the basics and obvious features of the watch. It is, however, directed at all of you who want to understand the nature of the watch and find the secret hidden treasures of the things it can do.

The bottom line is this. Think of the watch input and interaction in terms of this: the side button, where you power up, the crown which you can tap and rotate, and the screen which you can tap, swipe and Force Touch. Whatever screen you are on, if you can try each gesture between the three input methods, you will find that it will get you around and do things you wouldn't expect to. And one final input method is to invoke Siri.

It goes without saying that while the watch is powerful, it works even better when your friends have one too. The messaging between two watches gives you an additional dimension of messaging, like the digital touch and the heartbeat.

As far as notifications are concerned, the beeps and sounds, along with the haptic response, make the watch not only functional but also customizable and selective. You can mute messages but for a few important people, and you can create a digital touch. When you send them a touch, the haptic response in the recipient's watch is activated (obviously you can only do this if they have an Apple Watch too). It is a silent way to reach out and digitally touch a loved one. Or, if you want to take it even further, you can Force Touch the display to call the New Message composition. Here, you can place two fingers on the screen and that will show a beating heart that is following your heart's rhythm.

You will also be able to hear your own heartbeat. When you remove your fingers, the message is sent with your heartbeat to the recipient (again, they must have an Apple Watch too).

That was the final trick of the watch we have for you here. The best advice you can get is to go and play around with your watch and, don't worry, you won't break it. In the worst case scenario, just reset it and everything will be back to normal.